U.S. Department
of Transportation
**National Highway
Traffic Safety
Administration**

www.nhtsa.gov

DOT HS 811 527

November 2011

Occupant Restraint Use in 2010

Results From the
National Occupant Protection Use Survey
Controlled Intersection Study

1. Report No. DOT HS 811 527	2. Government Accession No.	3. Recipient's Catalog No.
4. Title and Subtitle Occupant Restraint Use in 2010: Results From the National Occupant Protection Use Survey Controlled Intersection Study		5. Report Date November 2011
		6. Performing Organization Code NVS-421
7. Author(s) **Timothy M. Pickrell** [*] **and Tony Jianqiang Ye** [†]		8. Performing Organization Report No.
9. Performing Organization Name Mathematical Analysis Division, National Center for Statistics and Analysis National Highway Traffic Safety Administration U.S. Department of Transportation 1200 New Jersey Avenue SE. Washington, DC 20590		10. Work Unit No. (TRAIS)
		11. Contract or Grant No.
12. Sponsoring Agency Name and Address Mathematical Analysis Division, National Center for Statistics and Analysis National Highway Traffic Safety Administration U.S. Department of Transportation 1200 New Jersey Avenue SE. Washington, DC 20590		13. Type of Report and Period Covered NHTSA Technical Report
		14. Sponsoring Agency Code

15. Supplementary Notes

[*] Mathematical Statistician, Mathematical Analysis Division, National Center for Statistics and Analysis, NHTSA
[†] Statistician, Bowhead Systems Management Inc., contractor working at NHTSA

Abstract

This report presents results from the 2010 National Occupant Protection Use Survey (NOPUS) Controlled Intersection Study. NOPUS is the only nationwide probability-based occupant restraint use survey. This survey is conducted annually by the National Center for Statistics and Analysis of the National Highway Traffic Safety Administration. The 2010 NOPUS found that seat belt use in rear seats increased significantly from 70 percent in 2009 to 74 percent in 2010. Significant increases in rear-seat belt use in 2010 occurred in the following passenger categories: male passengers, passengers 16 to 24 years old, passengers 25 to 69 years old, and passengers who were members of other races. Restraint use for all children from birth to 7 years old stood at 89 percent in 2010 as compared to 88 percent in 2009. The 2010 restraint use rates for children from birth to 1 year old, children 1 to 3 years old, and children 4 to 7 years old, are 99 percent, 94 percent, and 83 percent respectively. Child restraint use continued to be higher in the West than in the other regions.

17. Key Words Seat belt use, child restraint use, observational survey, demographics, rear-seat seat belt use, occupant protection, NOPUS, controlled intersections, premature graduation, car seat		18. Distribution Statement Document is available to the public from the National Technical Information Service www.ntis.gov.	
19. Security Classif. (of this report) Unclassified	20. Security Classif. (of this page) Unclassified	21. No. of Pages 30	22. Price

Form DOT F 1700.7 (8-72)   Reproduction of completed page authorized

TABLE of CONTENTS

TABLE of FIGURES

LIST OF TABLES

Executive Summary

The National Occupant Protection Use Survey (NOPUS) is the only nationwide probability-based survey of seat belt use (for occupants 8 and older in both front and rear seats), motorcycle helmet use, child restraint use (for children less than 8 years old), and driver electronic device use in the United States. It is conducted annually by the National Center for Statistics and Analysis of the National Highway Traffic Safety Administration. NOPUS is comprised of two sub-surveys: the Moving Traffic Survey and the Controlled Intersection (CI) Study.

In the CI study, occupants of passenger vehicles without commercial or government markings are observed from the roadside at intersections controlled by stop signs or stop lights. Only stopped vehicles are observed to allow ample time to collect a variety of information required by the survey. NOPUS derives its estimates of seat belt use in rear seats, child restraint use, driver electronic device use, and demographic characteristics of vehicle occupants from the CI study.

This report presents results of occupant restraint use from the 2010 National Occupant Protection Use Survey Controlled Intersection Study. NHTSA will publish driver electronic device use results in a separate report.

The following are some of the major findings from the 2010 NOPUS Controlled Intersection Study:

Rear Seats:
- Seat belt use in rear seats increased significantly from 70 percent in 2009 to 74 percent in 2010.
- The significant increase in seat belt use in rear seats in 2010 occurred in a number of passenger categories, including male passengers, passengers 16 to 24 years old, passengers 25 to 69 years old, and passengers who are members of other races.

Front Seats:
- Seat belt use continued to be lower among 16- to 24-year-olds than other age groups.
- Seat belt use continued to be lower among males than females.
- Seat belt use continued to be lower among black occupants than occupants of the other race groups.
- Seat belt use continued to be lower among drivers driving alone than among drivers with passengers.

Child Restraint Use:
- Restraint use for children from birth to 7 years old stood at 89 percent in 2010 as compared to 88 percent in 2009.
- In 2010, 99 percent of infants, 94 percent of children 1 to 3 years old, and 83 percent of children 4 to 7 years old were restrained, either appropriately or inappropriately.
- About 93 percent of children under 8 rode in the rear seats of vehicles in 2010.
- Child restraint use continued to be higher in the West than in the other regions in 2010.
- Restraint use for children driven by belted drivers continued to be higher than for those driven by unbelted drivers.

1. Introduction

The National Occupant Protection Use Survey is the only nationwide probability-based survey of seat belt use (for occupants 8 and older in both front and rear seats), motorcycle helmet use, child restraint use (for children less than 8 years old), and driver electronic device use in the United States. It is conducted annually by the National Center for Statistics and Analysis of the National Highway Traffic Safety Administration. NOPUS is comprised of two sub-surveys: Moving Traffic (MT) Survey and Controlled Intersection (CI) Study.

In the MT survey, the shoulder belt use data of front-seat occupants and helmet use data of motorcyclists is collected by observing passenger vehicle occupants either at the roadside or, in the case of expressways, while riding in a vehicle in traffic. NOPUS derives its major estimates of front-seat belt use and motorcycle helmet use from the MT survey. NHTSA published the results from the 2010 NOPUS MT survey in late 2010.[1][2] In contrast, the CI study data is collected at intersections controlled by stop signs or stoplights, where vehicle occupants are observed from the roadside. Only the stopped vehicles are observed due to time constraints restricting the amount of time available to collect the variety of information required by the survey. NOPUS derives its estimates of rear-seat belt use, child restraint use, driver electronic device use, and demographic characteristics of the vehicle occupants from the CI study.

Only motorcycles and passenger vehicles without commercial or government markings are observed by the NOPUS (NOPUS does not record restraint use data for occupants of commercial vehicles, buses, taxis, or emergency vehicles). The population of interest includes all 50 States, the District of Columbia, with the sample observation sites consisting of Federal, State, county highways, residential streets, and rural roads. Data is collected only during daylight hours when light is adequate to observe seat belt use through the vehicle windshield.

The 2010 NOPUS data collection was conducted between 7 a.m. and 6 p.m. during the period from June 7, 2010, to June 26, 2010. The 2010 NOPUS survey data is based on the results of 69,747 occupants observed in the 49,475 vehicles at the 1,446 data collection sites. Of these observed occupants, 3,914 were children under 8. More details on the NOPUS sampling, data collection and estimation are discussed in Section 5: NOPUS Methodology.

The purpose of this report is to present occupant restraint use results from the 2010 National Occupant Protection Use Survey Controlled Intersection Study. NHTSA will publish results on driver electronic device use in a separate report. In the years prior to 2009, NHTSA usually presented the results from the NOPUS CI study through three or four Research Notes, each of which covers one specific topic. This 2010 report, like its 2009 counterpart,[3] will pool together as much data as possible from the 2010 NOPUS CI study for the convenience of data users. However, in order to be consistent with the publications released by NHTSA in the years before 2009, sections in this report are arranged to cover similar topics to those in the Research Notes published previously.[4][5][6]

Please note that the terms "significant" and "statistically significant" are used interchangeably throughout this report. "Significant" always means "statistically significant" and the statistical significance level is chosen to be 0.1.

2. Demographic Results

Seat belt use nationwide was 85 percent in 2010, a slight gain from the 2009 use rate of 84 percent. This section presents the demographic breakdown of the occupants who used seat belts in 2010.

Although the NOPUS CI data is collected solely from vehicles stopped at intersections controlled by stop signs or stoplights, the estimates in this publication concerning seat belt use in the front seat reflect use by occupants in transit on all types of roadways. This is accomplished by making adjustments using data from the MT survey that observes seat belt use in vehicles in transit on general roadways.

Table 1 on page 5 presents results of passenger vehicle occupant seat belt use by demographic and other characteristics in 2009 and 2010, as well as the changes between the two years. Some major results are highlighted below.

Age

In 2010, there was no significant change in the seat belt use as compared to 2009 across all four age groups of occupants: 8 to 15 years old, 16 to 24 years old, 25 to 69 years old and 70 and older. Figure 1 shows a comparison of the seat belt use rates between 2009 and 2010 among occupants of these age groups.

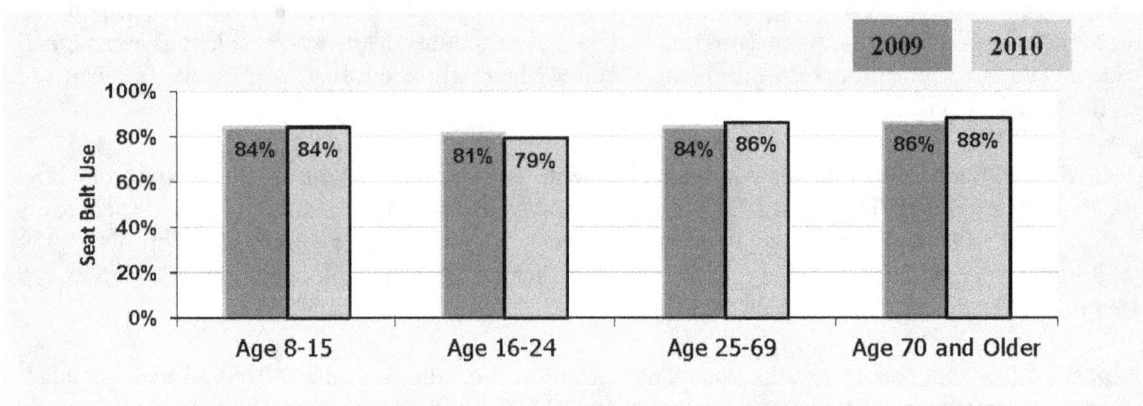

Figure 1: Seat Belt Use by Age for Occupants 8 and Older in 2009 and 2010

Figure 2 displays the trends of seat belt use for the four age groups over a period of 9 years (2002 to 2010). It shows that in 2010, seat belt use continued to be lower among 16- to 24-year-olds than other age groups.

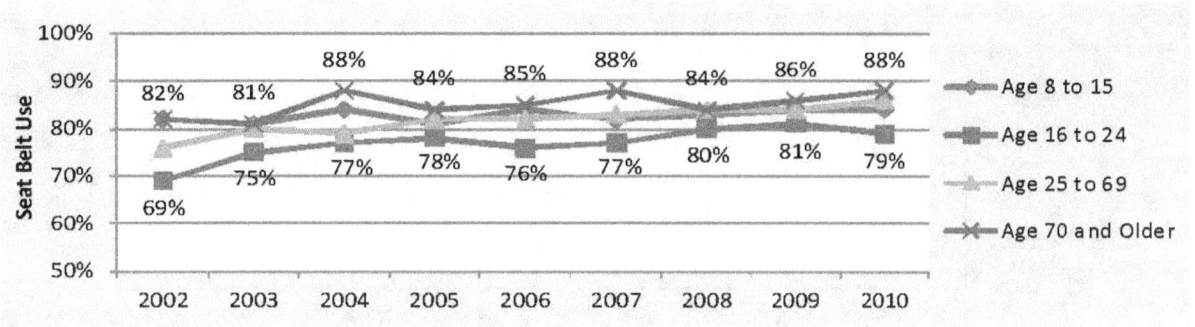

Figure 2: Seat Belt Use by Age for Occupants 8 and Older, 2002-2010

This figure also shows that in 2010 the seat belt use for occupants 70 and older was 88 percent, which is significantly higher than the other age groups.

Gender

Figure 3 shows the trends of seat belt use among male and female occupants over a period of 9 years (2002 to 2010). In 2010, seat belt use continued to be lower among males (83%) than females (88%).

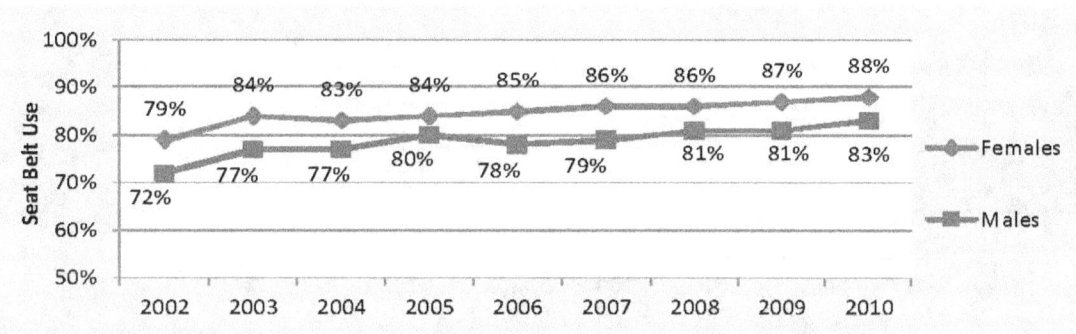

Figure 3: Seat Belt Use by Gender for Occupants 8 and Older, 2002-2010

Race

In NOPUS, the race of vehicle occupants is recorded as: black, white, and members of other races. The characterization is based on the visual assessment by the data collectors who observe vehicle occupants from roadsides.

Figure 4 shows the trends of seat belt use among occupants who are white, black, and members of other races over a period of 6 years (2005 to 2010). In 2010, seat belt use continued to be lower among black occupants than occupants of the other race groups. Seat belt use for members of other races was significantly higher than the complementary group (white and black occupants combined).

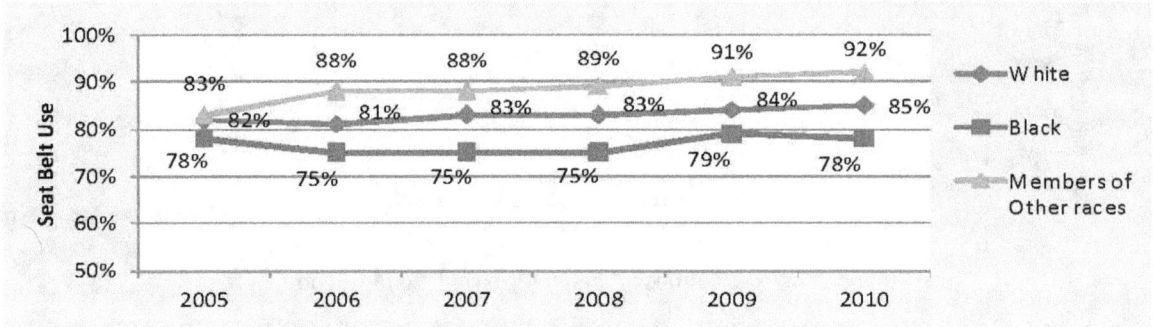

Figure 4: Seat Belt Use by Race for Occupants 8 and Older, 2005-2010

Presence of Passengers and Seat Belt Use

Figure 5 displays a clear pattern that seat belt use continued to be lower among drivers driving alone than among drivers driving with passengers.

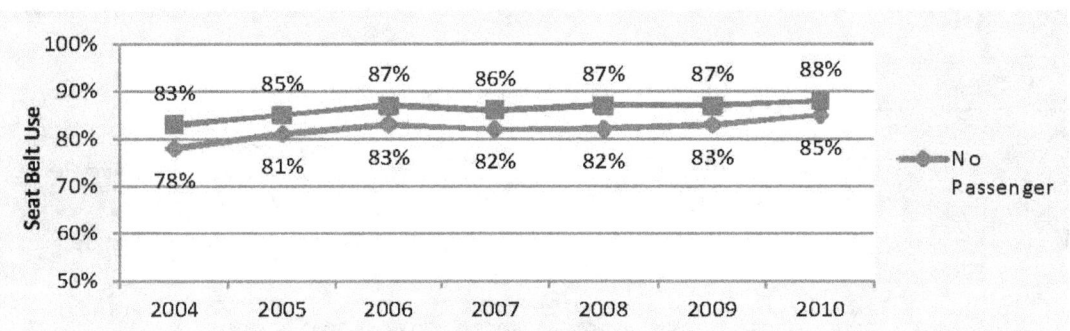

Figure 5: Passenger Effect on Seat Belt Use for Occupants 8 and Older, 2004-2010

Table 1: Passenger Vehicle Occupant Seat Belt Use by Demographic and Other Characteristics

Occupant Group[1]	2009 Belt Use[2]	2009 Confidence That Use Is High or Low in Group[3]	2010 Belt Use[2]	2010 Confidence That Use Is High or Low in Group[3]	2009-2010 Change in Percentage Points	2009-2010 Confidence in a Change in Percentage[4]
All Occupants	84%		85%		1	75%
Males[5]	81%	**100%**	83%	**100%**	2	78%
Females[5]	87%	**100%**	88%	**100%**	1	51%
Occupants by Age Group[5]						
Age 8 to 15	84%	58%	84%	74%	0	1%
Age 16 to 24	81%	**98%**	79%	**100%**	-2	52%
Age 25 to 69	84%	**90%**	86%	**100%**	2	87%
Age 70 and Older	86%	**94%**	88%	**100%**	2	86%
Occupants by Race[5]						
White	84%	75%	85%	54%	1	81%
Black	79%	**99%**	78%	**100%**	-1	29%
Members of Other Races	91%	**100%**	92%	**100%**	1	63%
Drivers With						
No Passengers	83%	**100%**	85%	**100%**	2	74%
At Least One Passenger	87%	**100%**	88%	**100%**	1	71%
Drivers With						
No Passengers	83%	**100%**	85%	**100%**	2	74%
Passengers All Under Age 8	88%	**100%**	89%	**100%**	1	50%
Passengers All Age 8 and Older	87%	**100%**	88%	**100%**	1	68%
Some Passengers Under Age 8 and Some Age 8 or Older	90%	**100%**	90%	**100%**	0	11%
Drivers Age 16-24 With						
No Passengers	83%	85%	79%	76%	-4	87%
Passengers All Age 16-24	80%	**91%**	77%	87%	-3	51%
At Least One Passenger Not Age 16-24	83%	51%	86%	**100%**	3	50%
Occupants Age 16-24 When						
All Occupants Are Age 16-24	81%	53%	78%	**100%**	-3	80%
At Least One Occupant Is Not Age 16-24	81%	53%	84%	**100%**	3	67%

[1] Drivers and right-front passengers of passenger vehicles with no commercial or government markings.
[2] Use of shoulder belts observed between 7 a.m. and 6 p.m.
[3] The statistical confidence that use in the occupant group (e.g., occupants who are members of other races) is higher or lower than use in the corresponding complementary occupant groups (e.g., combined black and white occupants). Confidences that meet or exceed 90 percent are formatted in boldface type. Confidences are rounded to the nearest percentage point, and so confidences reported as "100 percent" are between 99.5 percent and 100 percent.
[4] The degree of statistical confidence that the 2010 use rate is different from the 2009 rate. Confidences that meet or exceed 90 percent are formatted in boldface type.
[5] The age, gender, and racial classifications are based on the subjective assessments of roadside observers.
Source: NOPUS

3. Seat Belt Use in Rear Seats

Not all vehicles on the road today have shoulder belts in the rear seats. Based on the 2009 vehicle registration data from the National Vehicle Population Profile, R.L. Polk & Co., an estimated 90 percent of passenger vehicles on the road have shoulder belts in the rear outboard seating positions. Of the 10 percent of vehicles that have only lap belts in the rear outboard seats, all rear-seat vehicle occupants are counted by NOPUS as *not using shoulder belts*, regardless of whether they are using lap belts. Consequently, NOPUS rear-seat shoulder belt use estimates reflect both the degree to which vehicle occupants use restraints and the availability of shoulder belts in these seating positions.

Please note that rear-seat occupants might be underestimated in NOPUS because NOPUS only observes up to two passengers in the second row of seats and none in the third row and beyond.

Table 3 on page 8 presents results of seat belt use in the rear seat of passenger vehicles in 2009 and 2010 as well as the changes between the two years. Some major results are highlighted below.

Seat Belt Use in Rear Seats Increased Significantly in 2010

As shown in Figure 6, seat belt use in rear seats among all passengers 8 and older increased significantly from 70 percent in 2009 to 74 percent in 2010.

The significant increases in seat belt use in rear seats in 2010 also occurred in a number of passenger categories, including male passengers, passengers 16 to 24 years old, passengers 25 to 69 years old, and passengers who are members of other races. Figure 6 also shows the increases in these categories.

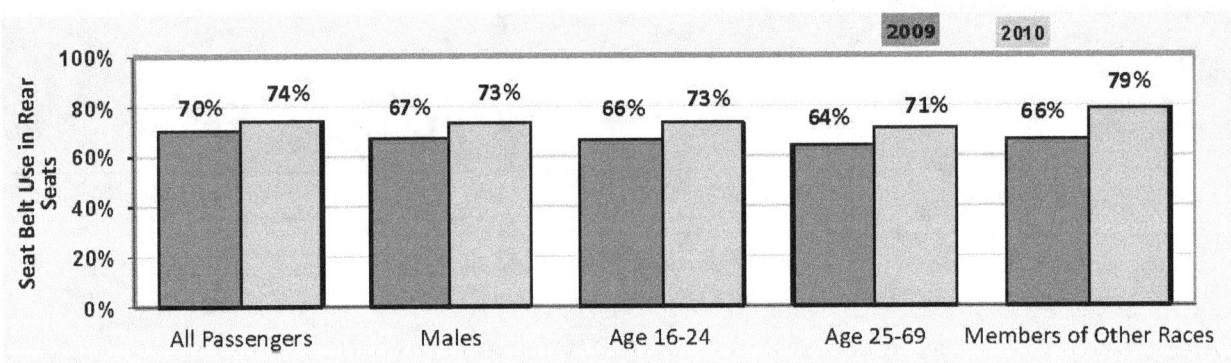

Figure 6: Seat Belt Use in Rear Seats Increased Significantly in 2010

Seat Belt Use in Rear Seats Versus in Front Seats

Figure 7 displays the trends of seat belt use in rear and front seats over a period of 7 years (2004 to 2010). It shows that, as in the previous years, seat belt use in 2010 was lower in the rear seat than in the front seat.

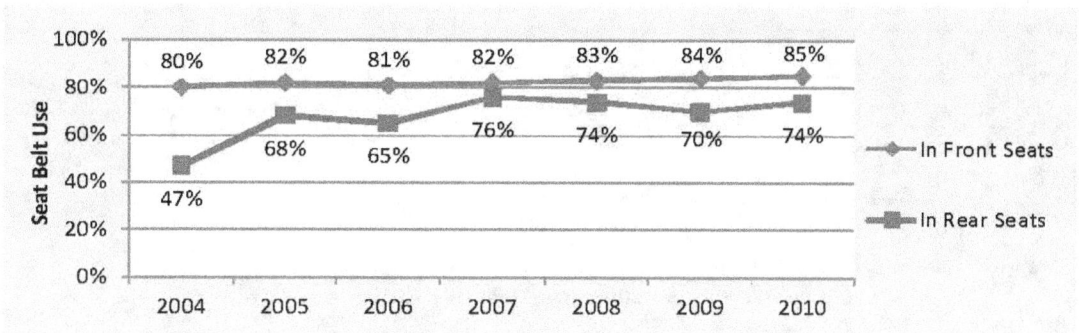

Figure 7: Seat Belt Use by Seating Position for Occupants 8 and Older, 2004-2010

State Laws and Rear-Seat Belt Use

At the time the 2010 NOPUS survey was conducted, 24 States and the District of Columbia required all vehicle occupants 18 and older to use seat belts when riding in the rear seat. Since July 1, 2009, rear-seat seat belt use laws have been enacted in Louisiana, Texas, New Jersey and Kansas. New Jersey's law took effect in January 2011. Table 2 provides a list of these States.

Table 2: States With Laws Requiring Seat Belts Be Used in All Seating Positions

Alaska	California	Delaware
District of Columbia	Idaho	Indiana
Kansas	Kentucky	Louisiana
Maine	Massachusetts	Minnesota
Montana	Nevada	New Mexico
North Carolina	Oregon	Rhode Island
South Carolina	Texas	Utah
Vermont	Washington	Wisconsin
Wyoming		

States with laws in effect as of June 30, 2010, requiring people 18 and older to use seat belts in all seating positions. Also includes the District of Columbia. The rear-seat seat belt use laws took effect in Louisiana, Texas, and Kansas during the period July 1, 2009 – June 30, 2010.

Figure 8 shows the trends of rear-seat belt use among passengers in the States with or without laws requiring belt use in all seating positions over a period of six years (2005 to 2010). As in the previous years, seat belt use in rear seats in 2010 was higher in the States with laws requiring belt use in all seating positions (79%) than in the States requiring belt use only in the front seat (69%).

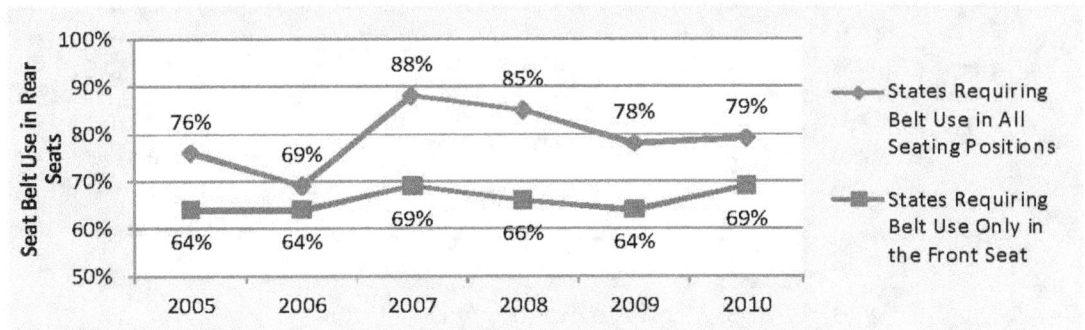

Figure 8: Seat Belt Use in Rear Seats by State Law Type for Occupants 8 and Older, 2005-2010

Table 3: Seat Belt Use in the Rear Seat of Passenger Vehicles, by Major Characteristics

Passenger Group[1]	2009		2010		2009-2010 Change	
	Belt Use[2]	Confidence That Use Is High or Low in Group[3]	Belt Use[2]	Confidence That Use Is High or Low in Group[3]	Change in Percentage Points	Confidence in a Change in Percentage[4]
All Passengers	70%		74%		4	**91%**
Males[5]	67%	**99%**	73%	78%	6	**96%**
Females[5]	72%	**99%**	74%	78%	2	59%
Passengers by Age Group [5]						
Age 8 to 15	76%	**100%**	75%	79%	-1	20%
Age 16 to 24	66%	**99%**	73%	57%	7	**97%**
Age 25 to 69	64%	**100%**	71%	**94%**	7	**92%**
Age 70 and Older	82%	**99%**	81%	**97%**	-1	11%
Passengers by Race [5]						
White	73%	**100%**	75%	**97%**	2	69%
Black	56%	**100%**	53%	**100%**	-3	34%
Members of Other Races	66%	81%	79%	**96%**	13	**98%**
Passengers in States With Laws Requiring Belts Be Used						
In All Seating Positions	78%	**100%**	79%	**99%**	1	71%
In the Front Seat Only	64%	**100%**	69%	**99%**	5	**91%**

[1] Up to two passengers observed in the second row of seats in passenger vehicles with no commercial or government markings.
[2] Use of shoulder belts observed between 7 a.m. and 6 p.m.
[3] The statistical confidence that use in the passenger group (e.g., passengers who are members of other races) is higher or lower than use in the corresponding complementary passenger groups (e.g., combined black and white passengers). Confidences that meet or exceed 90 percent are formatted in boldface type. Confidences are rounded to the nearest percentage point, and so confidences reported as "100 percent" are between 99.5 percent and 100.0 percent.
[4] The degree of statistical confidence that the 2010 use rate is different from the 2009 rate. Confidences that meet or exceed 90 percent are formatted in boldface type.
[5] The age, gender, and racial classifications are based on the subjective assessments of roadside observers.
Source: NOPUS

4. Child Restraint Use

In 2010, NOPUS continued to collect roadside observational data on child restraint use for all children under 8 years old. Table 5 on page 12 presents results of child restraint use in passenger motor vehicles by major characteristics in 2009 and 2010 as well as the changes between the two years. Table 7 on page 14 divides the occupants into three age groups and reports restraint use by some other characteristics among these groups. Table 6 on page 13 presents results on child rear placement by major characteristics in 2009 and 2010 as well as the changes between the two years. Some of the major results of child restraint use are discussed below.

Child Restraint Use among All Children Age Under 8

Figure 9 shows the trend of child restraint use since 2002. It shows that the restraint use among all children age under 8 stood at 89 percent in 2010 as compared to 88 percent in 2009.

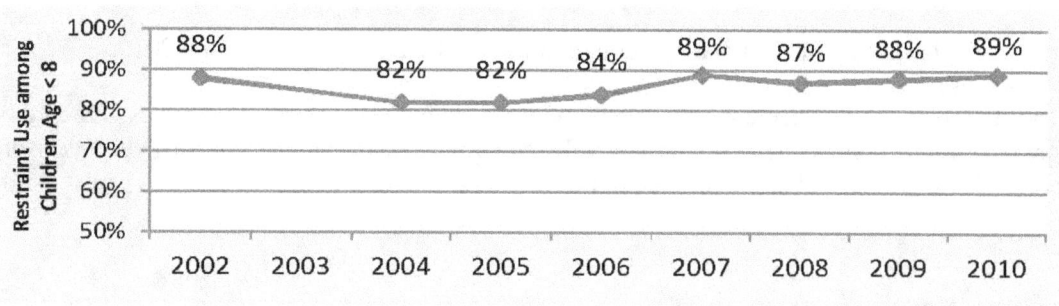

Figure 9: Child Restraint Use Among Children Under Age 8, 2002-2010

Child Restraint Use by Age

Before 2007, NHTSA published the child restraint use rates by age based on the data from the NOPUS. For the years 2007, 2008, and 2009, however, NHTSA's published estimates of child restraint use by age came from the National Survey of the Use of Booster Seats (NSUBS). As explained in NHTSA's reports and research notes in previous years,[3][6][7] since information about age is obtained by interviews in NSUBS and through visual assessment in NOPUS, the former is more accurate. For more details of the NSUBS, please refer to the NHTSA technical report "The 2009 National Survey of the Use of Booster Use,"[8] which is available at the Web site http://www-nrd.nhtsa.dot.gov/CMSWeb/index.aspx.

However, in 2010, no NSUBS was conducted. Thus, child restraint use rates from the 2010 NOPUS are reported here. Please note that since the rates are based on two different datasets, comparison of the 2009 and 2010 child restraint use rates by age should be avoided.

As shown in Figure 10, 99 percent of children from birth to 12 months, 94 percent of children 1 to 3 years old, and 83 percent children 4 to 7 years old were restrained in 2010.

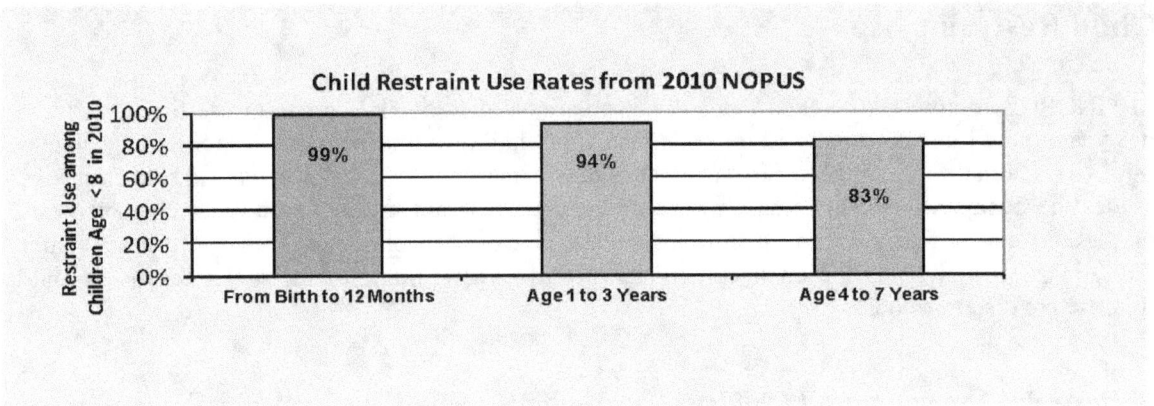

Figure 10: Child Restraint Use by Age (Data Source: 2010 NOPUS)

Child Rear Seat Placement

Figure 11 shows the trends of rear seat placement of children under 8 between 2002 and 2010.

The 2010 NOPUS found that 93 percent of children under 8 rode in the rear seats of vehicles. Of all the infants (from birth to 12 months), 97 percent rode in the rear seat. Ninety-nine percent of 1- to 3-year-olds and 89 percent of 4- to 7-year-old children were in the rear seats in 2010.

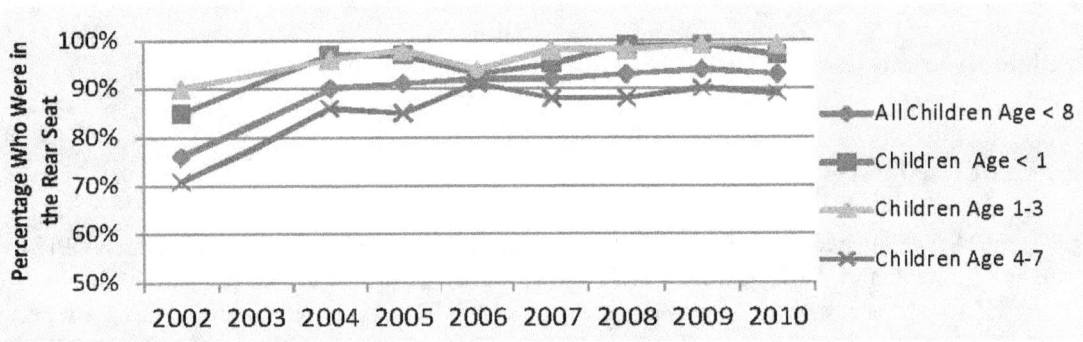

Figure 11: Child Rear Seat Placement, 2002 - 2010

At the time the 2010 survey was conducted, 9 States required children 5 and younger who weighed less than 80 pounds and were less than 54 inches tall to ride in the rear seats of vehicles. Table 4 lists the States with child rear placement laws.

Table 4: States With Laws Requiring Children 5 and Younger Be in the Rear Seat*

California	Georgia	Maine
New Jersey	Rhode Island	South Carolina
Tennessee	Washington	Wyoming

* Among children less than 80 pounds and less than 54" tall. States with laws in effect as of June 30, 2010. In no other States did such laws take effect during the period July 1, 2009, to June 30, 2010. In Delaware, children 11 and younger and 65 inches or less must be the rear seat if passenger air bag is active.

Child Restraint Use by Region

As shown in Figure 12, the child restraint use continued to be higher in the West than in the other regions in 2010.

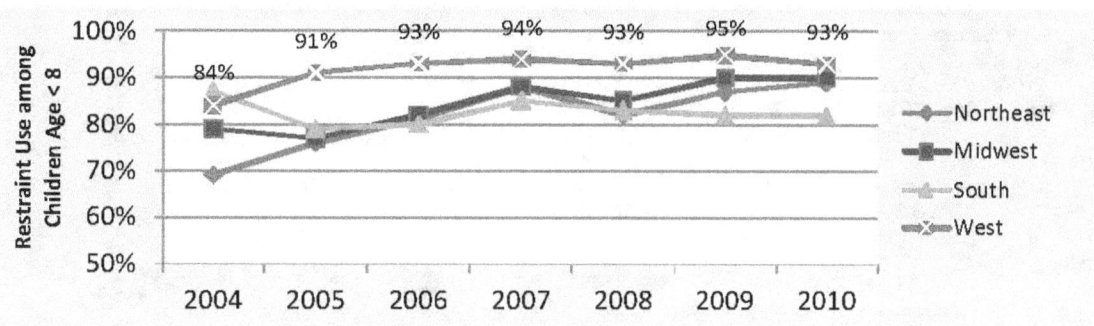

Figure 12: Child Restraint Use by Region, 2004-2010

Child Restraint Use by Driver Belt Status

As shown in Figure 13, the restraint use for children driven by belted drivers continued to be higher than for those driven by unbelted drivers.

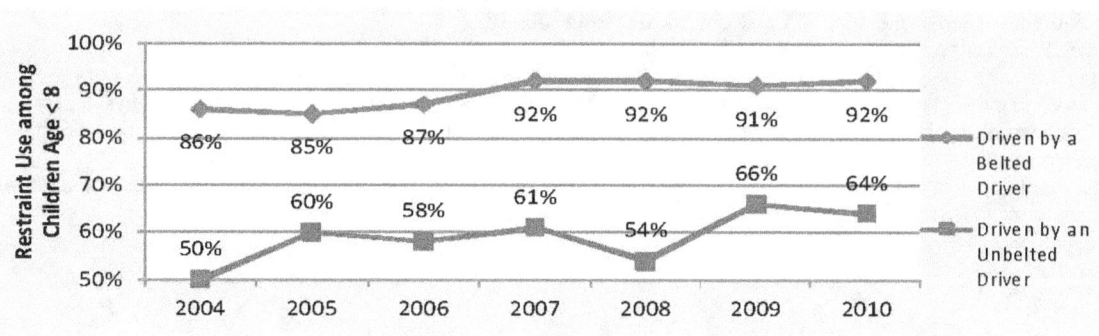

Figure 13: Child Restraint Use by Driver Belt Status, 2004-2010

Table 5: Child Restraint Use in Passenger Motor Vehicles, by Major Characteristics

Child Passenger Group[1]	2009		2010		2009–2010 Change	
	Restraint Use[2]	Confidence That Use Is High or Low in Group[3]	Restraint Use[2]	Confidence That Use Is High or Low in Group[3]	Change in Percentage Points	Confidence in a Change in Use[4]
All Child Passengers (From Birth to 7 Years)	88%		89%		1	49%
Children Driven by						
a Belted Driver	91%	**100%**	92%	**100%**	1	54%
an Unbelted Driver	66%	**100%**	64%	**100%**	-2	31%
a Male Driver	87%	79%	87%	**99%**	0	2%
a Female Driver	89%	79%	90%	**99%**	1	71%
a Driver Age 16 to 24	89%	66%	90%	65%	1	5%
a Driver Age 25 to 69	88%	63%	89%	55%	1	51%
a Driver Age 70 and Older	86%	64%	85%	70%	-1	7%
a White Driver	90%	**100%**	92%	**100%**	2	76%
a Black Driver	71%	**100%**	76%	**100%**	5	77%
a Driver who is a Member of Other Races	88%	50%	84%	**96%**	-4	60%
Children in						
the Front Seat	74%	**100%**	81%	**99%**	7	69%
the Rear Seat	89%	**100%**	90%	**99%**	1	42%
Child Passengers on						
Expressways	90%	89%	89%	64%	-1	10%
Surface Streets	87%	89%	89%	64%	2	63%
Child Passengers Traveling in						
Fast Traffic	90%	**93%**	90%	78%	0	10%
Medium-Speed Traffic	87%	82%	88%	85%	1	28%
Slow Traffic	87%	61%	90%	71%	3	53%
Child Passengers in						
Passenger Cars	84%	**100%**	86%	**100%**	2	85%
Vans & SUVs	93%	**100%**	92%	**100%**	-1	50%
Pickup Trucks	80%	**99%**	84%	**93%**	4	63%
Child Passengers in the						
Northeast	87%	61%	89%	51%	2	45%
Midwest	90%	83%	91%	85%	1	38%
South	82%	**100%**	82%	**99%**	0	1%
West	95%	**100%**	93%	**100%**	-2	45%
Child Passengers in						
Urban Areas	86%	79%	80%	**100%**	-6	89%
Suburban Areas	89%	80%	91%	**100%**	2	72%
Rural Areas	87%	67%	89%	51%	2	54%
Child Passengers Traveling During						
Weekdays	88%	61%	90%	**94%**	2	89%
Rush Hours	88%	50%	90%	55%	2	83%
Nonrush Hours	88%	50%	90%	55%	2	76%
Weekends	88%	61%	87%	**94%**	-1	49%

[1] Passengers under age 8 observed between 7 a.m. and 6 p.m. in the right-front seat or the second row of seats in passenger vehicles with no commercial or government markings that are stopped at a stop sign or stop light. Age, gender, and racial classifications are based on the subjective assessments of roadside observers.

[2] Use of child car seats (forward- or rear-facing), booster seats, and seat belts.

[3] The statistical confidence that use in the passenger group (e.g., child passengers in the Northeast) is higher or lower than use in the corresponding complementary passenger group (e.g., combined child passengers in the Midwest, in the South and in the West). Confidences that meet or exceed 90 percent are formatted in boldface type. Confidences are rounded to the nearest percentage point, and so confidences reported as "100 percent" are between 99.5 percent and 100.0 percent.

[4] The degree of statistical confidence that the 2010 use rate is different from the 2009 rate. Confidences that meet or exceed 90 percent are formatted in boldface type.

Source: NOPUS

Table 6: The Percent of Children Who Rode in the Rear Seat, by Major Characteristics

Child Passenger Group[1]	2009		2010		2009–2010 Change	
	Percentage Who Were in Rear Seat[2]	Confidence That Use Is High or Low in Group[3]	Percentage Who Were in Rear Seat[2]	Confidence That Use Is High or Low in Group[3]	Change in Percentage Points	Confidence in a Change in Rear Seat Occupancy[4]
All Child Passengers (From Birth to 7 Years)	94%		93%		-1	81%
Age 0 (Infants)	99%	100%	97%	96%	-2	57%
Age 1-3	99%	100%	99%	100%	0	58%
Age 4-7	90%	100%	89%	100%	-1	61%
Child Passengers in States With[5]						
Law Requiring Children From Birth of 5 Years Be in the Rear Seat	96%	85%	94%	89%	-2	60%
No Such Law	94%	85%	92%	89%	-2	80%
Children Driven by						
a Belted Driver	95%	99%	93%	93%	-2	88%
an Unbelted Driver	90%	99%	88%	93%	-2	22%
a Male Driver	94%	66%	93%	75%	-1	24%
a Female Driver	94%	66%	93%	75%	-1	89%
a Driver Age 16 to 24	98%	100%	98%	100%	0	27%
a Driver Age 25 to 69	94%	98%	92%	100%	-2	93%
a Driver Age 70 and Older	84%	94%	94%	62%	10	81%
a White Driver	94%	65%	93%	80%	-1	89%
a Black Driver	96%	92%	91%	77%	-5	90%
a Driver who is a Member of Other Races	92%	84%	95%	99%	3	75%
Child Passengers on						
Expressways	96%	100%	95%	100%	-1	57%
Surface Streets	93%	100%	92%	100%	-1	75%
Child Passengers Traveling in						
Fast Traffic	94%	54%	95%	99%	1	42%
Medium-Speed Traffic	94%	56%	92%	95%	-2	91%
Slow Traffic	94%	53%	92%	66%	-2	55%
Child Passengers in						
Passenger Cars	96%	100%	94%	86%	-2	87%
Vans & SUVs	95%	92%	93%	73%	-2	79%
Pickup Trucks	76%	100%	82%	100%	6	89%
Child Passengers in the						
Northeast	97%	100%	96%	99%	-1	45%
Midwest	92%	90%	92%	75%	0	17%
South	93%	76%	90%	96%	-3	83%
West	94%	57%	94%	83%	0	22%
Child Passengers in						
Urban Areas	96%	91%	94%	72%	-2	42%
Suburban Areas	95%	89%	93%	65%	-2	87%
Rural Areas	92%	96%	92%	88%	0	3%
Child Passengers Traveling During						
Weekdays	93%	97%	93%	61%	0	25%
Rush Hours	92%	88%	92%	91%	0	34%
Nonrush Hours	94%	88%	94%	91%	0	3%
Weekends	96%	97%	93%	61%	-3	85%
Child Passengers in a						
Rear-Facing Car Seat	99%	100%	100%	100%	1	51%
Forward-Facing Car Seat	99%	100%	99%	100%	0	52%
High-Backed Booster Seat	100%	UA	100%	100%	0	75%
Seat belt or Backless Booster Seat	90%	100%	86%	100%	-4	95%
No Restraint Observed	88%	100%	88%	98%	0	1%

[1] Passengers under 8 observed between 7 a.m. and 6 p.m. in the right-front seat or the second row of seats in passenger vehicles with no commercial or government markings that are stopped at a stop sign or stoplight. Age, gender, and racial classifications are based on the subjective assessments of roadside observers.

[2] The percentage of the child passenger group who were in the second row of seats at the time of observation.

[3] The statistical confidence that use in the passenger group (e.g., child passengers in the Northeast) is higher or lower than use in the corresponding complementary passenger group (e.g., combined child passengers in the Midwest, in the South and in the West). Confidences that meet or exceed 90 percent are formatted in boldface type. Confidences are rounded to the nearest percentage point, and so confidences reported as "100 percent" are between 99.5 percent and 100.0 percent.

[4] The degree of statistical confidence that the percentage of the child passenger group who were in the rear seat in 2010 is different from the analogous percentage from 2009.

[5] Use rates reflect the law in effect at the time data was collected.   UA: Estimate not available

Source: NOPUS

Table 7: Child Restraint Use in Passenger Motor Vehicles, by Age and Other Characteristics

Child Passenger Group[1]	2009		2010		2009–2010 Change	
	Restraint Use[2]	Confidence That Use Is High or Low in Group[3]	Restraint Use[2]	Confidence That Use Is High or Low in Group[3]	Change in Percentage Points	Confidence in a Change in Use[4]
Infants (From Birth to 12 Months)						
Infants Driven by						
a Belted Driver	99%	**97%**	99%	84%	0	13%
an Unbelted Driver	93%	**97%**	94%	84%	1	23%
a Male Driver	98%	79%	97%	**98%**	-1	30%
a Female Driver	99%	79%	100%	**98%**	1	**94%**
Infants in						
Passenger Cars	98%	75%	99%	86%	1	**93%**
Vans & SUVs	100%	UA	99%	56%	-1	82%
Pickup Trucks	NA	NA	NA	NA	NA	NA
Infants in the						
Northeast	98%	61%	97%	87%	-1	19%
Midwest	96%	**93%**	100%	**99%**	4	**95%**
South	99%	82%	99%	68%	0	23%
West	100%	UA	99%	69%	-1	68%
Infants in						
Urban Areas	99%	87%	99%	63%	0	18%
Suburban Areas	99%	81%	99%	61%	0	6%
Rural Areas	97%	89%	99%	57%	2	79%
Children Age 1 to 3						
Children Age 1-3 Driven by						
a Belted Driver	94%	**99%**	95%	**99%**	1	42%
an Unbelted Driver	85%	**99%**	80%	**99%**	-5	51%
a Male Driver	95%	70%	93%	89%	-2	55%
a Female Driver	93%	70%	95%	89%	2	68%
Children Age 1-3 in						
Passenger Cars	92%	**97%**	92%	**99%**	0	5%
Vans & SUVs	97%	**100%**	96%	**99%**	-1	34%
Pickup Trucks	83%	**96%**	94%	56%	11	**90%**
Children Age 1-3 in the						
Northeast	96%	81%	94%	57%	-2	31%
Midwest	93%	68%	96%	**93%**	3	77%
South	89%	**98%**	89%	**93%**	0	5%
West	97%	**98%**	96%	**93%**	-1	36%
Children Age 1-3 in						
Urban Areas	91%	90%	86%	**95%**	-5	52%
Suburban Areas	95%	90%	96%	**93%**	1	36%
Rural Areas	93%	70%	95%	65%	2	54%
Children Age 4 to 7						
Children Age 4-7 Driven by						
a Belted Driver	86%	**100%**	87%	**100%**	1	41%
an Unbelted Driver	54%	**100%**	51%	**100%**	-3	20%
a Male Driver	81%	83%	82%	89%	1	35%
a Female Driver	83%	83%	85%	89%	2	37%
Children Age 4-7 in						
Passenger Cars	75%	**100%**	79%	**100%**	4	81%
Vans & SUVs	90%	**100%**	88%	**100%**	-2	37%
Pickup Trucks	79%	78%	79%	83%	0	8%
Children Age 4-7 in the						
Northeast	80%	72%	83%	52%	3	58%
Midwest	86%	**92%**	87%	**92%**	1	18%
South	74%	**99%**	74%	**100%**	0	2%
West	92%	**100%**	89%	**99%**	-3	54%
Children Age 4-7 in						
Urban Areas	80%	70%	72%	**99%**	-8	74%
Suburban Areas	83%	65%	86%	**96%**	3	65%
Rural Areas	82%	53%	84%	66%	2	43%

[1] Passengers under 8 observed between 7 a.m. and 6 p.m. in the right-front seat or the second row of seats in passenger vehicles with no commercial or government markings that are stopped at a stop sign or stoplight. Age, gender, and racial classifications are based on the subjective assessments of roadside observers.
[2] Use of child car seats (forward- or rear-facing), booster seats, and seat belts.
[3] The statistical confidences that use in the passenger group (e.g., child passengers in the Northeast) is higher or lower than use in the corresponding complementary passenger group (e.g., combined child passengers in the Midwest, in the South and in the West). Confidences that meet or exceed 90 percent are formatted in boldface type. Confidences are rounded to the nearest percentage point, and so confidences reported as "100 percent" are between 99.5 percent and 100.0 percent.
[4] The degree of statistical confidence that the 2010 use rate is different from the 2009 rate. Confidences that meet or exceed 90 percent are formatted in boldface type.
NA: Data not sufficient to produce a reliable estimate.   UA: Estimate not available.   Source: NOPUS

5. NOPUS Methodology

This section briefly discusses the sample design, data collection, and estimation used in the 2010 NOPUS Controlled Intersection Study. For more details on the NOPUS methodology, please refer to the upcoming NHTSA technical report, "National Occupant Protection Use Survey Methodology," which will be available at http://www-nrd.nhtsa.dot.gov/Cats/index.aspx. Data collection, estimation, and variance estimation for NOPUS are conducted by Westat, Inc., under the direction of NHTSA's National Center for Statistics and Analysis under Federal contract number DTNH22-07-D-00057.

Sample Design

NOPUS is in the final stage of its phasing-in to the redesigned sample of observational sites. The redesigned sample contains a greater proportion of local roads, new roads and thus provides a better representation of the current U.S. population of motor vehicle occupants. Since the initial implementation of the redesign in 2006 and every year following, NOPUS has had an increasing number of sites from the redesigned sample and a decreasing number of sites from the old sample. Data from 2005 and prior years were obtained from the old observational sites only.

The old sample design was a multistage, stratified sample design in which 50 primary sampling units (PSUs) were selected at the first stage and the strata were created using Census region, metropolitan statistical area (MSA) or non-MSA status, and the State's seat belt use rate (high = 70% and over, medium = 55 to under 70% and low = under 55%). In the second stage, roads were classified into two secondary strata prior to sampling, namely major roads (including limited access highways, U.S. roads, and State routes), and local roads (including county roads, residential roads, and rural roads). Major roads were then sampled from computerized road inventories supplied by State DOT offices in the 25 States represented by the sampled PSUs while local roads were clustered within Census tracts in the selected PSUs. For the 2010 NOPUS, 12 PSUs and 449 sites were selected from the old design.

The redesigned NOPUS sample was selected using a two-stage design with stratified probability proportional to size (PPS) sampling at each stage. The sampling frame of PSUs for the 2006 redesigned sample included all counties in the U.S. but excluded Puerto Rico and the U.S. Territories. In the redesigned sample, only one PSU was designated as a certainty sampling unit (i.e., probability one) due to its large vehicle miles traveled (VMT). In order to decrease the variances associated with the survey estimates, the remaining PSUs were stratified according to their predicted rates of restraint use based on a regression model that used primary enforcement law status, ratio of fatal crashes to VMT, percentage of college graduates, and several other relevant variables as predictors. The non-certainty PSUs were selected by systematic PPS sampling from these primary strata using VMT as the measure of size. The secondary sampling units (SSUs) consisted of road segments that lie at least partly inside the selected PSUs. To define road segments, the selected PSUs were divided into grids, usually of one-acre in size. For the 2010 NOPUS, 43 PSUs and 1,334 sites were selected from the redesigned sample.

Table 9 shows the observed sample sizes of the 2010 NOPUS. A total of 69,747 occupants were observed in the 48,331 vehicles at the 1,446 data collection sites. Of these observed occupants, 3,914 were children under 8. Please note that due to ineligibility, construction, danger in the area, or road closure, observations could not be completed at some of the sampled observation sites.

Table 8: Sites, Vehicles and Occupants in the 2010 NOPUS

Numbers of	2009	2010	Percentage Change
Sites Observed	1,496	1,446	-3%
Vehicles Observed	49,475	48,331	-2%
Occupants Age 8 and Older	66,950	65,833	-2%
In Front Seat	63,682	62,349	-2%
In Rear Seat	3,268	3,484	7%
Occupants Under Age 8	3,543	3,914	10%
Children Under Age 1	392	540	38%
Children Age 1 to 3	1,225	1,317	8%
Children Age 4 to 7	1,926	2,057	7%

Data Collection

The 2010 NOPUS data collection was conducted during the period from June 7, 2010 to June 26, 2010.

In the NOPUS Controlled Intersection Study, trained data collectors observe *restraint use* of drivers and other occupants of passenger vehicles having no commercial or government markings which have stopped at a stop sign or stoplight during daylight hours between 7 a.m. and 6 p.m. Observations are made both on the surface streets and at the ends of the expressway exit ramps (where there are controlled intersections.) Only stopped vehicles are observed based on the time required to collect the variety of information required by the survey, including subjective assessments of the vehicle occupants' age and race. Observers collect data on the driver, right-front passenger, and up to two passengers in the second row of seats. Observers do not interview vehicle occupants intentionally, allowing NOPUS to capture the uninfluenced behavior of the occupants.

NOPUS Controlled Intersection Study is always done following NOPUS Moving Traffic Survey and is usually scheduled for all surface streets and limited access highway ramps, where NOPUS data from previous years indicates that a controlled intersection exists. If the data collectors arrive at an assigned surface street site and the site is not controlled, they are instructed to search for an alternative. The data collectors move down the roadside and record vehicle and occupant characteristics. Once the traffic light turns green or they finish observing all vehicles, the data collectors return to the intersection to wait for the next traffic light cycle or next vehicle. They observe vehicles in the lane closest to their observational position, even if the closest lane is an exclusive turn lane (which is often the case at the controlled intersections.) When possible and if visibility allows, the data collectors also observe the other lanes of traffic. The data collectors are instructed to record the first behavior of the driver in which they observe.

Regardless of road type, the data collectors observe vehicles at the assigned intersections for 40 minutes. Since data collection for the CI study immediately follows the MT survey, no additional vehicle counts are conducted at controlled intersections. Instead, the independent counts from the MT survey observation sites are used for the corresponding CI study sites.

Estimation

NOPUS estimates the rate of occupants restrained in restraint type (R) among the occupants having characteristic (C) using the formula,

$$\text{Restraint Use}_{CR} = \frac{\sum_{i,j,k} w_{ijk} F_{ijk} CR_{ijk}}{\sum_{i,j,k} w_{ijk} F_{ijk} C_{ijk}},$$

where w_{ijk} and F_{ijk}, respectively, denote the base weight and the product of various weight adjustment factors at the site k in the stratum j of the PSU i. CR_{ijk} stands for the number of observed occupants having characteristic C and restrained in restraint type R and C_{ijk} denotes the number of observed occupants having characteristic C at the site k in the stratum j of the PSU i. For example, the seat belt use by vehicle type is estimated using the above formula, where CR_{ijk} is the number of observed belted occupants in certain type of vehicles (such as passenger cars, vans & SUVs, or pickup trucks) and C_{ijk} is the number of ALL (belted and unbelted) occupants observed in that type of vehicles at the site k in the stratum j of the PSU i.

In certain instances, NHTSA does not provide estimates. These are typically restraint use estimates whose numerator is based on fewer than five persons observed, whose denominator is based on fewer than 30 people observed, or the estimates are not statistically different from 0% (i.e., the standard error is at least half the point estimate). These are reported as "NA" in publications. Any related estimate (i.e., change in use and confidence estimates) is not reported as well. The same criteria are used in reporting estimates from the NSUBS survey.

6. References

[1] Pickrell, T. M., & Ye, T. J. (2010, September). *Seat Belt Use in 2010 – Overall Results*, (DOT HS 811 378). Washington, DC: National Highway Traffic Safety Administration. Available at www-nrd.nhtsa.dot.gov/Pubs/811378

[2] Pickrell, T. M., & Ye, T. J. (2010, December). *Motorcycle Helmet Use in 2010 – Overall Results*. (DOT HS 811 419). Washington, DC: National Highway Traffic Safety Administration. Available at www-nrd.nhtsa.dot.gov/Pubs/811419

[3] Pickrell, T.M., & Ye, T.J. (2010, November). *Occupant Restraint Use in 2009 – Results from the National Occupant Protection Use Survey Controlled Intersection Study*. (DOT HS 811 414). Washington, DC: National Highway Traffic Safety Administration. Available at www-nrd.nhtsa.dot.gov/Pubs/811414

[4] Pickrell, T. M., & Ye, T. J. (2009, August). *Seat Belt Use in 2008 – Demographic Results*. (DOT HS 811 183). Washington, DC: National Highway Traffic Safety Administration. Available at www-nrd.nhtsa.dot.gov/Pubs/811183

[5] Pickrell, T. M., & Ye, T. J. (2009, May). *Seat Belt Use in Rear Seats in 2008*. (DOT HS 811 133). Washington, DC: National Highway Traffic Safety Administration. Available at www-nrd.nhtsa.dot.gov/Pubs/811133

[6] Pickrell, T. M., & Ye, T. J. (2009, May). *Child Restraint Use in 2008 – Overall Results*. (DOT HS 811 135). Washington, DC: National Highway Traffic Safety Administration. Available at www-nrd.nhtsa.dot.gov/Pubs/811135

[7] Ye, T. J., & Pickrell, T. M. (2008, April). *Child Restraint Use in 2007 – Overall Results*. (DOT HS 810 931). Washington, DC: National Highway Traffic Safety Administration. Available at www-nrd.nhtsa.dot.gov/Pubs/810931

[8] Pickrell, T. M., & Ye, T. J. (2010, September). *The 2009 National Survey of Use of Booster Seats*. (DOT HS 811 377). Washington, DC: National Highway Traffic Safety Administration. Available at www-nrd.nhtsa.dot.gov/Pubs/811377

Appendix: Definitions

- Vehicle occupants observed in the NOPUS survey are counted as "belted" if they appeared to have a shoulder belt across the front of the body. NOPUS does not observe the use of lap belts because these restraints cannot be reliably observed from the roadside.

- The survey classifies a child as:
    - <u>Restrained in a rear-facing car seat</u> if the child appears to be on a seat on top of the vehicle seat, facing the rear of the vehicle, with harness straps across the front of the child.
    - <u>Restrained in a forward-facing car seat</u> if the child appears to be on a seat on top of the vehicle seat, facing the front of the vehicle, with harness straps across the front of the child.
    - <u>Restrained in a high-backed booster seat</u> if the child appears to be on a seat on top of the vehicle seat with a shoulder belt across the front of the child.
    - <u>Restrained in a seat belt or backless booster seat</u> if there is a shoulder belt across the front of the child but the observers cannot see if the child is in a seat on top of the vehicle seat.
    - <u>Restrained</u> if s/he is restrained by any of the above.
    - The remaining children are classified as <u>unrestrained</u>. Note that in the survey there is no mention of being "unrestrained" in, for example, a forward-facing car seat. NOPUS does not observe the use of lap belts, and does not distinguish between seat belts and backless booster seats, because these assessments cannot be reliable if observed from the roadside.

- The racial categories "Black," "White," and "Members of other races" in NOPUS reflect subjective characterizations by roadside observers regarding the race of vehicle occupants. Likewise observers record all age groups (8 to 15 years old, 16 to 24 years old, 25 to 69 years old, and 70 and older) that best fits their visual assessment of each observed occupant.

- "Expressways" are defined as roadways with limited access, while "surface streets" comprise all other roadways.

- A roadway is defined to have "fast traffic" if, during the observation period, the average speed of passenger vehicles passing the observer(s) exceeds 50 mph, with "medium-speed traffic" defined as 31 to 50 mph and "slow traffic" defined as 30 mph or slower. The traffic speed data in the CI survey are matched to the MT survey data.

- A roadway is defined to have "heavy traffic" if the average number of vehicles per lane mile on the roadway during the observation period exceeded 45 vehicles per lane mile. "Moderately dense traffic" is defined as 26 to 45 vehicles per lane per mile. "Light traffic" is defined as a maximum of 25 vehicles per lane per mile. The traffic density data in the CI survey is matched to the MT survey data.

- Since NOPUS is not a census but based on some probability sample, it is impossible to produce State-by-State restraint use results. However NOPUS can and does produce regional estimates using the following categories:

<u>Northeast</u>: Connecticut, Massachusetts, Maine, New Hampshire, New Jersey, New York, Pennsylvania, Rhode Island, Vermont

<u>Midwest</u>: Iowa, Kansas, Illinois, Indiana, Michigan, Minnesota, Missouri, North Dakota, Nebraska, Ohio, South Dakota, Wisconsin

<u>South</u> : Alabama, Arkansas, the District of Columbia, Delaware, Florida, Georgia, Kentucky, Louisiana, Maryland, Mississippi, North Carolina, Oklahoma, South Carolina, Tennessee, Texas, Virginia, West Virginia

 <u>West</u>: Alaska, Arizona, California, Colorado, Hawaii, Idaho, Montana, New Mexico, Nevada, Oregon, Washington, Washington, Wyoming

These definitions of the four NOPUS regions are the same regional definitions used in the NSUBS.